Chapter 9
Relate Fractions and Decimals

Houghton
Mifflin
Harcourt

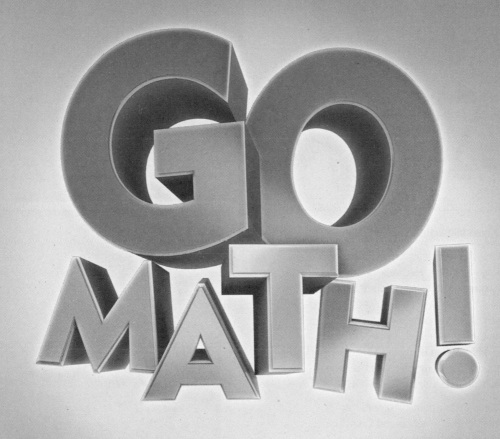

GO MATH!

© Houghton Mifflin Harcourt Publishing Company • Cover Image Credits: (Ground Squirrel) ©Don Johnston/All Canada Photos/Getty Images; (Sawtooth Range. Idaho) ©Ron and Patty Thomas Photography/E+/Getty Images

Houghton
Mifflin
Harcourt

Printed in the U.S.A.

ISBN 978-0-544-34227-9

14 0877 22 21 20 19 18

4500695342   D E F G

Dear Students and Families,

Welcome to **Go Math!**, Grade 4! In this exciting mathematics program, there are hands-on activities to do and real-world problems to solve. Best of all, you will write your ideas and answers right in your book. In **Go Math!**, writing and drawing on the pages helps you think deeply about what you are learning, and you will really understand math!

By the way, all of the pages in your **Go Math!** book are made using recycled paper. We wanted you to know that you can Go Green with **Go Math!**

Sincerely,

The Authors

Made in the United States
Text printed on 100% recycled paper

© Houghton Mifflin Harcourt Publishing Company • Image Credits: (bg) ©Sankar Salvady/Flickr/Getty Images; (t) ©Blaine Harrington III/Alamy Images; (c) ©Don Johnston/All Canada Photos/Getty Images; (b) ©Erich Kuchling/Westend61/Corbis

GO MATH!

Authors

Juli K. Dixon, Ph.D.
Professor, Mathematics Education
University of Central Florida
Orlando, Florida

Edward B. Burger, Ph.D.
President, Southwestern University
Georgetown, Texas

Steven J. Leinwand
Principal Research Analyst
American Institutes for
 Research (AIR)
Washington, D.C.

Matthew R. Larson, Ph.D.
K-12 Curriculum Specialist for
 Mathematics
Lincoln Public Schools
Lincoln, Nebraska

Martha E. Sandoval-Martinez
Math Instructor
El Camino College
Torrance, California

Contributor

Rena Petrello
Professor, Mathematics
Moorpark College
Moorpark, California

English Language Learners Consultant

Elizabeth Jiménez
CEO, GEMAS Consulting
Professional Expert on English
 Learner Education
Bilingual Education and
 Dual Language
Pomona, California

Fractions and Decimals

 Critical Area Developing an understanding of fraction equivalence, addition and subtraction of fractions with like denominators, and multiplication of fractions by whole numbers

9 Relate Fractions and Decimals 493

COMMON CORE STATE STANDARDS

4.NF Number and Operations–Fractions
Cluster C Understand decimal notation for fractions, and compare decimal fractions.
4.NF.C.5, 4.NF.C.6, 4.NF.C.7

4.MD Measurement and Data
Cluster A Solve problems involving measurement and conversion of measurements from a larger unit to a smaller unit.
4.MD.A.2

GO DIGITAL

Go online! Your math lessons are interactive. Use *i*Tools, Animated Math Models, the Multimedia *e*Glossary, and more.

Chapter 9 Overview

In this chapter, you will explore and discover answers to the following **Essential Questions**:

- How can you record decimal notation for fractions, and compare decimal fractions?
- Why can you record tenths and hundredths as decimals and fractions?
- What are some different models you can use to find equivalent fractions?
- How can you compare decimal fractions?

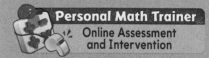

Personal Math Trainer
Online Assessment and Intervention

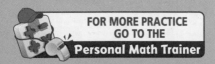
CRITICAL AREA REVIEW PROJECT FUNDRAISER: *www.thinkcentral.com*

FOR MORE PRACTICE
GO TO THE
Personal Math Trainer

Practice and Homework

Lesson Check and
Spiral Review in
every lesson

Relate Fractions and Decimals

 Show What You Know

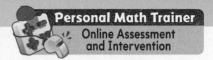
Check your understanding of important skills.

Name _____

▶ **Count Coins** **Find the total value.** (2.MD.C.8)

1.

Total value: _____

2.

Total value: _____

▶ **Equivalent Fractions**

Write two equivalent fractions for the picture. (3.NF.A.3b)

3.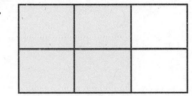

_____

4.

_____

▶ **Fractions with Denominators of 10**

Write a fraction for the words. You may draw a picture. (3.NF.A.1)

5. three tenths _____

6. six tenths _____

7. eight tenths _____

8. nine tenths _____

Math in the Real World

The Hudson River Science Barge, docked near New York City, provides a demonstration of how renewable energy can be used to produce food for large cities. Vegetables grown on the barge require _____ of the water needed by field crops. Use these clues to find the fraction and decimal for the missing amount.

- The number is less than one and has two decimal places.
- The digit in the hundredths place has a value of $\frac{5}{100}$.
- The digit in the tenths place has a value of $\frac{2}{10}$.

Chapter 9   **493**

Vocabulary Builder

▶ **Visualize It** ....

Complete the Semantic Map by using words with a ✓.

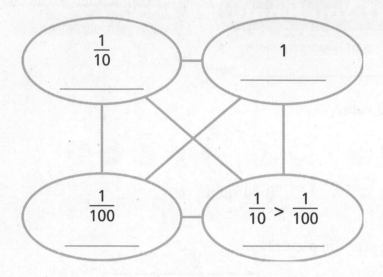

Review Words

✓ compare

equivalent fractions

fraction

place value

✓ whole

Preview Words

decimal

decimal point

equivalent decimals

✓ hundredth

✓ tenth

▶ **Understand Vocabulary** ....

Draw a line to match each word with its definition.

Word	Definition
1. decimal	• Two or more decimals that name the same amount
2. decimal point	• One part out of one hundred equal parts
3. tenth	• A number with one or more digits to the right of the decimal point
4. hundredth	• One part out of ten equal parts
5. equivalent decimals	• A symbol used to separate dollars from cents in money amounts and to separate the ones and the tenths places in decimals

GO DIGITAL
• Interactive Student Edition
• Multimedia eGlossary

Chapter 9 Vocabulary

compare

comparar

14

decimal

decimal

19

decimal point

punto decimal

20

equivalent decimals

decimales equivalentes

28

equivalent fractions

fracciones equivalentes

29

hundredth

centésimo

40

tenth

décimo

89

whole

entero

96

A number with one or more digits to the right of the decimal point

Examples: 0.5, 0.06, and 12.679 are decimals.

To describe whether numbers are equal to, less than, or greater than each other

Ones	.	Tenths	Hundredths
1	.	1	5
1	.	3	

1.3 > 1.15

Two or more decimals that name the same amount

Ones	.	Tenths	Hundredths
0	.	8	
0	.	8	0

Example: 0.8 and 0.80 are equivalent decimals.

A symbol used to separate dollars from cents in money amounts, and to separate the ones and the tenths places in a decimal

Example: 6.4
↑ decimal point

One of one hundred equal parts

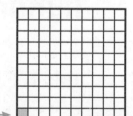

hundredth →

Two or more fractions that name the same amount

Example: $\frac{3}{4}$ and $\frac{6}{8}$ name the same amount.

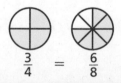

$$\frac{3}{4} = \frac{6}{8}$$

All of the parts of a shape or group

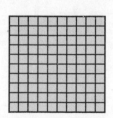

One of ten equal parts

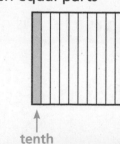

tenth

Matchup

For 2 to 3 players

Materials

- 1 set of word cards

How to Play

1. Put the cards face-down in rows. Take turns to play.

2. Choose two cards and turn them face-up.

   - If the cards show a word and its meaning, it's a match. Keep the pair and take another turn.

   - If the cards do not match, turn them back over.

3. The game is over when all cards have been matched. The players count their pairs. The player with the most pairs wins.

Word Box

compare

decimal

decimal point

equivalent
 decimals

equivalent
 fractions

hundredth

tenth

whole

Journal

The Write Way

Materials

Choose one idea. Write about it.

- Write a paragraph that uses at three of these words or phrases.

  decimal   decimal point   hundredth   tenth   whole

- Explain in your own words what equivalent decimals are.

- As the writer of a math advice column, you often hear from readers about their struggles to relate fraction and decimal forms. Write a letter explaining what those readers need to know to better understand this subject.

Name _____

Relate Tenths and Decimals

Essential Question How can you record tenths as fractions and decimals?

Common Core
Number and Operations—
Fractions—4.NF.C.6
MATHEMATICAL PRACTICES
MP2, MP5, MP6

🔑 Unlock the Problem

Ty is reading a book about metamorphic rocks. He has read $\frac{7}{10}$ of the book. What decimal describes the part of the book Ty has read?

A **decimal** is a number with one or more digits to the right of the **decimal point**. You can write tenths and hundredths as fractions or decimals.

🔒 One Way Use a model and a place-value chart.

Fraction

Shade $\frac{7}{10}$ of the model.

Think: The model is divided into 10 equal parts. Each part represents one **tenth**.

Decimal

$\frac{7}{10}$ is 7 tenths.

Ones	.	Tenths	Hundredths
	.		

↑ decimal point

Write: _____

Write: _____

Read: seven tenths

Read: _____

🔒 Another Way Use a number line.

Label the number line with decimals that are equivalent to the fractions. Locate the point $\frac{7}{10}$.

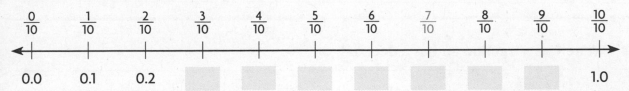

$\frac{0}{10}$ $\frac{1}{10}$ $\frac{2}{10}$ $\frac{3}{10}$ $\frac{4}{10}$ $\frac{5}{10}$ $\frac{6}{10}$ $\frac{7}{10}$ $\frac{8}{10}$ $\frac{9}{10}$ $\frac{10}{10}$

0.0   0.1   0.2   1.0

_____ names the same amount as $\frac{7}{10}$.

So, Ty read 0.7 of the book.

Math Talk

MATHEMATICAL PRACTICES ②

Use Reasoning How is the size of one whole related to the size of one tenth?

• How can you write 0.1 as a fraction? Explain.

_____

Tara rode her bicycle $1\frac{6}{10}$ miles. What decimal describes how far she rode her bicycle?

You have already written a fraction as a decimal. You can also write a mixed number as a decimal.

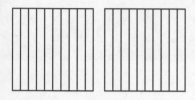 One Way Use a model and a place-value chart.

Fraction

Shade $1\frac{6}{10}$ of the model.

Write: _____

Read: one and six tenths

Decimal

$1\frac{6}{10}$ is 1 whole and 6 tenths.

Think: Use the ones place to record wholes.

Ones	.	Tenths	Hundredths
	.		

Write: _____

Read: _____

Another Way Use a number line.

Label the number line with equivalent mixed numbers and decimals. Locate the point $1\frac{6}{10}$.

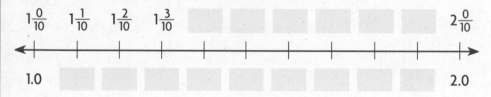

$1\frac{0}{10}$ $1\frac{1}{10}$ $1\frac{2}{10}$ $1\frac{3}{10}$ … $2\frac{0}{10}$

1.0 … 2.0

_____ names the same amount as $1\frac{6}{10}$.

So, Tara rode her bicycle _____ miles.

Try This! Write 1 as a fraction and as a decimal.

Shade the model to show 1.

Fraction: _____

Think: 1 is 1 whole and 0 tenths.

Ones	.	Tenths	Hundredths
	.		

Decimal: _____

Name _____

1. Write five tenths as a fraction and as a decimal.

   Fraction: _____   Decimal: _____

Ones	.	Tenths	Hundredths
	.		

Write the fraction or mixed number and the decimal shown by the model.

2.

_____   _____

3.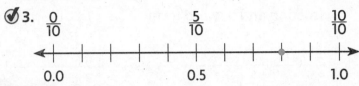

$\dfrac{0}{10}$ $\dfrac{5}{10}$ $\dfrac{10}{10}$

0.0   0.5   1.0

_____   _____

Math Talk

MATHEMATICAL PRACTICES ⑥

Attend to Precision How can you write $1\frac{3}{10}$ as a decimal? Explain.

Write the fraction or mixed number and the decimal shown by the model.

4.

_____   _____

5. $1\frac{0}{10}$ $1\frac{5}{10}$

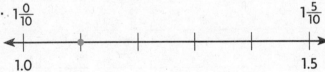

1.0   1.5

_____   _____

6.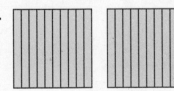

_____   _____

7. $3\frac{0}{10}$ $3\frac{5}{10}$ $4\frac{0}{10}$

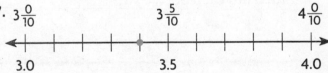

3.0   3.5   4.0

_____   _____

Practice: Copy and Solve Write the fraction or mixed number as a decimal.

8. $5\frac{9}{10}$

9. $\frac{1}{10}$

10. $\frac{7}{10}$

11. $8\frac{9}{10}$

12. $\frac{6}{10}$

13. $6\frac{3}{10}$

14. $\frac{5}{10}$

15. $9\frac{7}{10}$

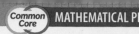
Problem Solving • Applications

Use the table for 16–19.

Ramon's Rock Collection	
Name	**Type**
Basalt	Igneous
Rhyolite	Igneous
Granite	Igneous
Peridotite	Igneous
Scoria	Igneous
Shale	Sedimentary
Limestone	Sedimentary
Sandstone	Sedimentary
Mica	Metamorphic
Slate	Metamorphic

16. What part of the rocks listed in the table are igneous? Write your answer as a decimal.

_____

17. Sedimentary rocks make up what part of Ramon's collection? Write your answer as a fraction and in word form.

_____

18. **THINK SMARTER** What part of the rocks listed in the table are metamorphic? Write your answer as a fraction and as a decimal.

_____

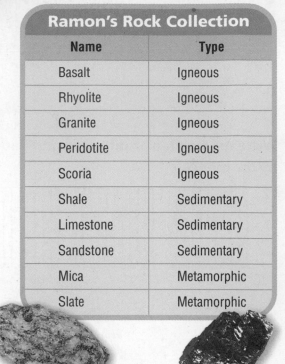
▲ Granite– Igneous   ▲ Mica–Metamorphic

▲ Sandstone– Sedimentary

19. **MATHEMATICAL PRACTICE 5 Communicate** Niki wrote the following sentence in her report: "Metamorphic rocks make up 2.0 of Ramon's rock collection." Describe her error.

_____

20. **GO DEEPER** Josh paid for three books with two $20 bills. He received $1 in change. Each book was the same price. How much did each book cost?

_____

21. **THINK SMARTER** Select a number shown by the model. Mark all that apply.

$1\frac{7}{10}$	$\frac{70}{10}$	1.7
7	0.7	$\frac{17}{10}$

Relate Tenths and Decimals

Common Core

COMMON CORE STANDARD—4.NF.C.6
Understand decimal notation for fractions, and compare decimal fractions.

Write the fraction or mixed number and the decimal shown by the model.

1. 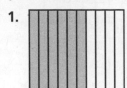 Think: The model is divided into 10 equal parts. Each part represents one tenth.

$\frac{6}{10}$; 0.6

2.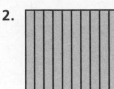

_____

3.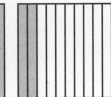
$2\frac{0}{10}$ $2\frac{5}{10}$

2.0   2.5

_____

Write the fraction or mixed number as a decimal.

4. $\frac{4}{10}$ 5. $3\frac{1}{10}$ 6. $\frac{7}{10}$ 7. $6\frac{5}{10}$ 8. $\frac{9}{10}$

_____   _____   _____   _____   _____

Problem Solving *Real World*

9. There are 10 sports balls in the equipment closet. Three are kickballs. Write the portion of the balls that are kickballs as a fraction, as a decimal, and in word form.

10. Peyton has 2 pizzas. Each pizza is cut into 10 equal slices. She and her friends eat 14 slices. What part of the pizzas did they eat? Write your answer as a decimal.

_____   _____

11. **WRITE** *Math* Do 0.3 and 3.0 have the same value? Explain.

_____

_____

Lesson Check (4.NF.C.6)

1. Valerie has 10 CDs in her music case. Seven of the CDs are pop music CDs. What is this amount written as a decimal?

2. What decimal amount is modeled below?

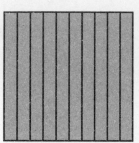 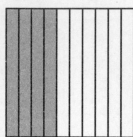

_____

Spiral Review (4.OA.B.4, 4.NF.A.1, 4.NF.B.3b)

3. Write one number that is a factor of 13.

4. An art gallery has 18 paintings and 4 photographs displayed in equal rows on a wall, with the same number of each type of art in each row. What could be the number of rows?

_____

5. How do you write the mixed number shown as a fraction greater than 1?

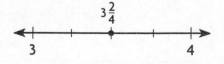

$3\frac{2}{4}$

3   4

6. What fraction of this model, in simplest form, is shaded?

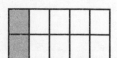

FOR MORE PRACTICE GO TO THE Personal Math Trainer

Name _____

Relate Hundredths and Decimals

Essential Question How can you record hundredths as fractions and decimals?

 Common Core **Number and Operations—Fractions—4.NF.C.6**
MATHEMATICAL PRACTICES
MP1, MP2, MP7

Unlock the Problem

In the 2008 Summer Olympic Games, the winning time in the men's 100-meter butterfly race was only $\frac{1}{100}$ second faster than the second-place time. What decimal represents this fraction of a second?

You can write hundredths as fractions or decimals.

• Circle the numbers you need to use.

One Way Use a model and a place-value chart.

Fraction

Shade $\frac{1}{100}$ of the model.

Think: The model is divided into 100 equal parts. Each part represents one **hundredth**.

Write: _____

Read: one hundredth

Decimal

Complete the place-value chart. $\frac{1}{100}$ is 1 hundredth.

Ones	.	Tenths	Hundredths
0	.	0	1

Write: _____

Read: one hundredth

Another Way Use a number line.

Label the number line with equivalent decimals. Locate the point $\frac{1}{100}$.

Math Talk MATHEMATICAL PRACTICES ❷

Use Reasoning How is the size of one tenth related to the size of one hundredth?

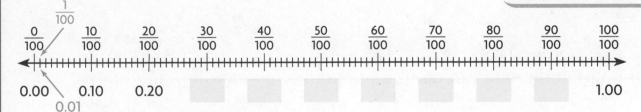

_____ names the same amount as $\frac{1}{100}$.

So, the winning time was _____ second faster.

Alicia won her 400-meter freestyle race by $4\frac{25}{100}$ seconds. How can you write this mixed number as a decimal?

One Way Use a model and a place-value chart.

Mixed Number

Shade the model to show $4\frac{25}{100}$.

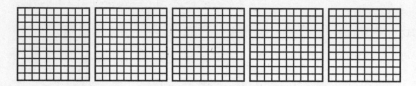

Write: _____

Read: four and twenty-five hundredths

Decimal

Complete the place-value chart.

Think: Look at the model above. $4\frac{25}{100}$ is 4 wholes and 2 tenths 5 hundredths.

Ones	.	Tenths	Hundredths
	.		

Write: _____

Read: _____

Another Way Use a number line.

Label the number line with equivalent mixed numbers and decimals. Locate the point $4\frac{25}{100}$.

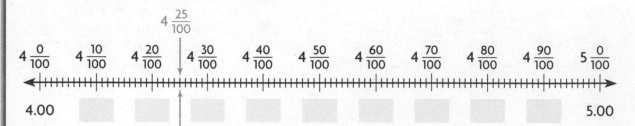

$4\frac{25}{100}$

$4\frac{0}{100}$ $4\frac{10}{100}$ $4\frac{20}{100}$ $4\frac{30}{100}$ $4\frac{40}{100}$ $4\frac{50}{100}$ $4\frac{60}{100}$ $4\frac{70}{100}$ $4\frac{80}{100}$ $4\frac{90}{100}$ $5\frac{0}{100}$

4.00   5.00

_____ names the same amount as $4\frac{25}{100}$.

So, Alicia won her race by _____ seconds.

Name _____

1. Shade the model to show $\frac{31}{100}$.

   Write the amount as a decimal. _____

Ones	.	Tenths	Hundredths
	.		

Write the fraction or mixed number and the decimal shown by the model.

2. _____

3. _____

4. $6\frac{0}{100}$ $6\frac{50}{100}$ $7\frac{0}{100}$

   6.00   6.50   7.00

_____ _____

On Your Own

Write the fraction or mixed number and the decimal shown by the model.

5. _____

6. _____

7. $\frac{0}{100}$ $\frac{50}{100}$ $\frac{100}{100}$

   0.00   0.50   1.00

_____ _____

Practice: Copy and Solve Write the fraction or mixed number as a decimal.

8. $\frac{9}{100}$ 9. $4\frac{55}{100}$ 10. $\frac{10}{100}$ 11. $9\frac{33}{100}$ 12. $\frac{92}{100}$ 13. $14\frac{16}{100}$

Problem Solving · Applications Real World

14. THINKSMARTER Shade the grids to show three different ways to represent $\frac{16}{100}$ using models.

15. MATHEMATICAL PRACTICE ① **Describe Relationships**
Describe how one whole, one tenth, and one hundredth are related.

_____

_____

16. THINKSMARTER Shade the model to show $1\frac{24}{100}$. Then write the mixed number in decimal form.

17. GO DEEPER The Memorial Library is 0.3 mile from school. Whose statement makes sense? Whose statement is nonsense? Explain your reasoning.

I am going to walk 3 tenths mile to the Memorial Library after school.

I am going to walk 3 miles to the Memorial Library after school.

Gabe	**Tara**
_____	_____
_____	_____
_____	_____

Name _____

Relate Hundredths and Decimals

Common Core

COMMON CORE STANDARD—4.NF.C.6
Understand decimal notation for fractions, and compare decimal fractions.

Write the fraction or mixed number and the decimal shown by the model.

1. Think: The whole is divided into one hundred equal parts, so each part is one hundredth.

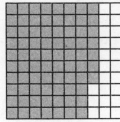

$\frac{77}{100}$; 0.77

2.

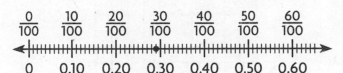

_____

3.

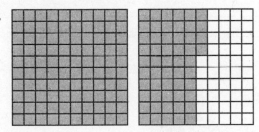

_____

4.

$4\frac{20}{100}$ $4\frac{30}{100}$ $4\frac{40}{100}$ $4\frac{50}{100}$ $4\frac{60}{100}$ $4\frac{70}{100}$ $4\frac{80}{100}$

4.20 4.30 4.40 4.50 4.60 4.70 4.80

_____

Write the fraction or mixed number as a decimal.

5. $\frac{37}{100}$

6. $8\frac{11}{100}$

7. $\frac{98}{100}$

8. $25\frac{50}{100}$

9. $\frac{6}{100}$

_____ _____ _____ _____ _____

Problem Solving

10. There are 100 pennies in a dollar. What fraction of a dollar is 61 pennies? Write it as a fraction, as a decimal, and in word form.

_____

11. **WRITE** *Math* Describe a situation where it is easier to use decimals than fractions, and explain why.

_____

Lesson Check (4.NF.C.6)

1. What decimal represents the shaded section of the model below?

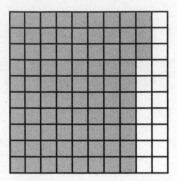

2. There were 100 questions on the unit test. Alondra answered 97 of the questions correctly. What decimal represents the fraction of questions Alondra answered correctly?

Spiral Review (4.OA.C.5, 4.NF.B.3b, 4.NF.B.3d, 4.NF.B.4c)

3. Write an expression that is equivalent to $\frac{7}{8}$.

4. What is $\frac{9}{10} - \frac{6}{10}$?

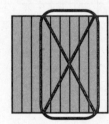

5. Misha used $\frac{1}{4}$ of a carton of 12 eggs to make an omelet. How many eggs did she use?

6. Kurt used the rule *add* 4, *subtract* 1 to generate a pattern. The first term in his pattern is 5. Write a number that could be in Kurt's pattern.

FOR MORE PRACTICE
GO TO THE
Personal Math Trainer

Equivalent Fractions and Decimals

Essential Question How can you record tenths and hundredths as fractions and decimals?

Common Core
Number and Operations—Fractions—4.NF.C.5 Also *4.NF.C.6*
MATHEMATICAL PRACTICES
MP2, MP6, MP7

Unlock the Problem

Daniel spent a day hiking through a wildlife preserve. During the first hour of the hike, he drank $\frac{6}{10}$ liter of water. How many hundredths of a liter did he drink?

- Underline what you need to find.
- How can you represent hundredths?

_____

One Way Write $\frac{6}{10}$ as an equivalent fraction with a denominator of 100.

MODEL

RECORD

$$\frac{6}{10} = \frac{6 \times }{10 \times } = \frac{}{100}$$

$$\frac{6}{10} = \frac{}{100}$$

Another Way Write $\frac{6}{10}$ as a decimal.

Think: 6 tenths is the same as 6 tenths 0 hundredths.

Ones	.	Tenths	Hundredths

So, Daniel drank _____, or _____ liter of water.

Math Talk

MATHEMATICAL PRACTICES ⑥

Explain how you can write 0.2 as hundredths.

- **Explain** why 6 tenths is equivalent to 60 hundredths.

_____

_____

Jasmine collected 0.30 liter of water in a jar during a rainstorm. How many tenths of a liter did she collect?

Equivalent decimals are decimals that name the same amount. You can write 0.30 as a decimal that names tenths.

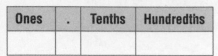 **One Way** Write 0.30 as an equivalent decimal.

Show 0.30 in the place-value chart.

Ones	.	Tenths	Hundredths

Think: There are no hundredths.

0.30 is equivalent to _____ tenths.

Write 0.30 as _____.

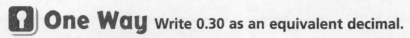 **Another Way** Write 0.30 as a fraction with a denominator of 10.

STEP 1 Write 0.30 as a fraction.

0.30 is _____ hundredths.

30 hundredths written as a fraction is _____.

_____

STEP 2 Write $\frac{30}{100}$ as an equivalent fraction with a denominator of 10.

Think: 10 is a common factor of the numerator and the denominator.

$$\frac{30}{100} = \frac{30 \div \boxed{}}{100 \div \boxed{}} = \frac{\boxed{}}{10}$$

So, Jasmine collected _____, or _____ liter of water.

Share and Show

1. Write $\frac{4}{10}$ as hundredths.

Write $\frac{4}{10}$ as an equivalent fraction.

$$\frac{4}{10} = \frac{4 \times \boxed{}}{10 \times \boxed{}} = \frac{\boxed{}}{100}$$

Fraction: _____

Write $\frac{4}{10}$ as a decimal.

Ones	.	Tenths	Hundredths
	.		

Decimal: _____

Write the number as hundredths in fraction form and decimal form.

2. $\frac{7}{10}$

3. 0.5

4. $\frac{3}{10}$

Write the number as tenths in fraction form and decimal form.

5. 0.40

6. $\frac{80}{100}$

7. $\frac{20}{100}$

On Your Own

Practice: Copy and Solve Write the number as hundredths in fraction form and decimal form.

Math Talk

MATHEMATICAL PRACTICES ②

Reason Abstractly Explain whether you can write 0.25 as tenths.

8. $\frac{8}{10}$

9. $\frac{2}{10}$

10. 0.1

Practice: Copy and Solve Write the number as tenths in fraction form and decimal form.

11. $\frac{60}{100}$

12. $\frac{90}{100}$

13. 0.70

THINKSMARTER Write the number as an equivalent mixed number with hundredths.

14. $1\frac{4}{10}$

15. $3\frac{5}{10}$

16. $2\frac{9}{10}$

Problem Solving • Applications (Real World)

17. **THINK SMARTER** Carter says that 0.08 is equivalent to $\frac{8}{10}$. Describe and correct Carter's error.

_____

_____

_____

18. **THINK SMARTER** For numbers 18a–18e, choose True or False for the statement.

18a. 0.6 is equivalent to $\frac{6}{100}$. ○ True ○ False

18b. $\frac{3}{10}$ is equivalent to 0.30. ○ True ○ False

18c. $\frac{40}{100}$ is equivalent to $\frac{4}{10}$. ○ True ○ False

18d. 0.40 is equivalent to $\frac{4}{100}$. ○ True ○ False

18e. 0.5 is equivalent to 0.50. ○ True ○ False

Connect to Science

Inland Water

How many lakes and rivers does your state have? The U.S. Geological Survey defines inland water as water that is surrounded by land. The Atlantic Ocean, the Pacific Ocean, and the Great Lakes are not considered inland water.

19. **WRITE** ▸Math Just over $\frac{2}{100}$ of the entire United States is inland water. Write $\frac{2}{100}$ as a decimal.

_____

20. **MATHEMATICAL PRACTICE 6** Can you write 0.02 as tenths? **Explain.**

_____

_____

21. About 0.17 of the area of Rhode Island is inland water. Write 0.17 as a fraction.

_____

22. **GO DEEPER** Louisiana's lakes and rivers cover about $\frac{1}{10}$ of the state. Write $\frac{1}{10}$ as hundredths in words, fraction form, and decimal form.

_____

Equivalent Fractions and Decimals

Common
Core

COMMON CORE STANDARD—4.NF.C.5
*Understand decimal notation for fractions,
and compare decimal fractions.*

Write the number as hundredths in fraction form and decimal form.

1. $\frac{5}{10}$

$$\frac{5}{10} = \frac{5 \times 10}{10 \times 10} = \frac{50}{100}$$

Think: 5 tenths is the same as 5 tenths and 0 hundredths. Write 0.50.

$$\frac{50}{100}; 0.50$$
_____

2. $\frac{9}{10}$

3. 0.2

4. 0.8

_____   _____   _____

Write the number as tenths in fraction form and decimal form.

5. $\frac{40}{100}$

6. $\frac{10}{100}$

7. 0.60

_____   _____   _____

Problem Solving · Real World

8. Billy walks $\frac{6}{10}$ mile to school each day. Write $\frac{6}{10}$ as hundredths in fraction form and in decimal form.

_____

9. **WRITE** *Math* Write $\frac{5}{10}$ in three equivalent forms.

_____

_____

_____

_____

Lesson Check (4.NF.C.5)

1. The fourth-grade students at Harvest School make up 0.3 of all students at the school. What fraction is equivalent to 0.3?

2. Kyle and his brother have a marble set. Of the marbles, 12 are blue. This represents $\frac{50}{100}$ of all the marbles. What decimal is equivalent to $\frac{50}{100}$?

Spiral Review (4.OA.C.5, 4.NF.A.1, 4.NF.B.4c, 4.NF.C.6)

3. Jesse won his race by $3\frac{45}{100}$ seconds. What is this number written as a decimal?

4. Marge cut 16 pieces of tape for mounting pictures on poster board. Each piece of tape was $\frac{3}{8}$ inch long. How much tape did Marge use?

5. Of Katie's pattern blocks, $\frac{9}{12}$ are triangles. What is $\frac{9}{12}$ in simplest form?

6. A number pattern has 75 as its first term. The rule for the pattern is *subtract* 6. What is the sixth term?

FOR MORE PRACTICE GO TO THE
Personal Math Trainer

Name _____

Relate Fractions, Decimals, and Money

Essential Question How can you relate fractions, decimals, and money?

 Common Core **Number and Operations—Fractions—4.NF.C.6**
MATHEMATICAL PRACTICES
MP2, MP4, MP6

🔑 Unlock the Problem

Together, Julie and Sarah have $1.00 in quarters. They want to share the quarters equally. How many quarters should each girl get? How much money is this?

> **Remember**
> 1 dollar = 100 cents
> 1 quarter = 25 cents
> 1 dime = 10 cents
> 1 penny = 1 cent

🔓 Use the model to relate money, fractions, and decimals.

4 quarters = 1 dollar = $1.00

$0.25 $0.25 $0.25 $0.25

1 quarter is $\frac{25}{100}$, or $\frac{1}{4}$ of a dollar.

2 quarters are $\frac{50}{100}$, $\frac{2}{4}$, or $\frac{1}{2}$ of a dollar.

$\frac{1}{2}$ of a dollar = $0.50, or 50 cents.

Circle the number of quarters each girl should get.

So, each girl should get 2 quarters, or $ _____ .

🔓 Examples Use money to model decimals.

1 dollar	10 dimes = 1 dollar	100 pennies = 1 dollar

$1.00, or	1 dime = $\frac{1}{10}$, or 0.10 of a dollar	1 penny = $\frac{1}{100}$, or 0.01 of a dollar
_____ cents	$ _____ , or 10 cents	$ _____ , or 1 cent

Math Talk

MATHEMATICAL PRACTICES ④

Model Mathematics Model 68 pennies. What part of a dollar do you have? Explain.

Relate Money and Decimals Think of dollars as ones, dimes as tenths, and pennies as hundredths.

$1.56

Dollars	.	Dimes	Pennies
1	.	5	6

Think: $1.56 = 1 dollar and 56 pennies

There are 100 pennies in 1 dollar.
So, $1.56 = 156 pennies.

1.56 dollars

Ones	.	Tenths	Hundredths
1	.	5	6

Think: 1.56 = 1 one and 56 hundredths

There are 100 hundredths in 1 one.
So, 1.56 = 156 hundredths.

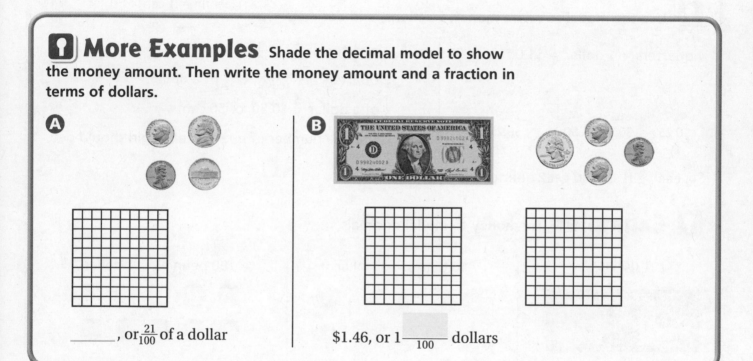

🔓 **More Examples** Shade the decimal model to show the money amount. Then write the money amount and a fraction in terms of dollars.

A

_____, or $\frac{21}{100}$ of a dollar

B

$1.46, or $1\frac{}{100}$ dollars

Try This! Complete the table to show how money, fractions, mixed numbers, and decimals are related.

$ Bills and Coins	Money Amount	Fraction or Mixed Number	Decimal
	$0.03		0.03
	$0.25	$\frac{25}{100}$, or $\frac{1}{4}$	
2 quarters 1 dime		$\frac{60}{100}$, or $\frac{6}{10}$	
2 $1 bills 5 nickels			

Math Talk

MATHEMATICAL PRACTICES ❷

Reason Abstractly Would you rather have $0.25 or $\frac{3}{10}$ of a dollar? Explain.

Name _____

Share and Show

1. Write the amount of money as a decimal in terms of dollars.

   5 pennies = $\frac{5}{100}$ of a dollar = _____ of a dollar.

Write the total money amount. Then write the amount as a fraction or a mixed number and as a decimal in terms of dollars.

2.

   _____ _____

✔3.

   _____ _____

Write as a money amount and as a decimal in terms of dollars.

4. $\frac{92}{100}$ _____

5. $\frac{7}{100}$ _____

6. $\frac{16}{100}$ _____

✔7. $\frac{53}{100}$ _____

Math Talk

MATHEMATICAL PRACTICES ❻

Make Connections How are $0.84 and $\frac{84}{100}$ of a dollar related?

On Your Own

Write the total money amount. Then write the amount as a fraction or a mixed number and as a decimal in terms of dollars.

8.

   _____ _____

9.

   _____ _____

Write as a money amount and as a decimal in terms of dollars.

10. $\frac{27}{100}$ _____

11. $\frac{4}{100}$ _____

12. $\frac{75}{100}$ _____

13. $\frac{100}{100}$ _____

Write the total money amount. Then write the amount as a fraction and as a decimal in terms of dollars.

14. 1 quarter 6 dimes 8 pennies

    _____ _____

15. 3 dimes 5 nickels 20 pennies

    _____ _____

MATHEMATICAL PRACTICE 6 **Make Connections** **Algebra** **Complete to tell the value of each digit.**

16. $1.05 = _____ dollar + _____ pennies,   1.05 = _____ one + _____ hundredths

17. $5.18 = _____ dollars + _____ dime + _____ pennies

 5.18 = _____ ones + _____ tenth + _____ hundredths

Problem Solving • Applications

Use the table for 18–19.

18. The table shows the coins three students have. Write Nick's total amount as a fraction in terms of dollars.

_____

Pocket Change				
Name	Quarters	Dimes	Nickels	Pennies
Kim	1	3	2	3
Tony	0	6	1	6
Nick	2	4	0	2

19. **THINK SMARTER** Kim spent $\frac{40}{100}$ of a dollar on a snack. Write as a money amount the amount she has left.

_____

20. **GO DEEPER** Travis has $\frac{1}{2}$ of a dollar. He has at least two different types of coins in his pocket. Draw two possible sets of coins that Travis could have.

21. **THINK SMARTER** Complete the table.

$ Bills and Coins	Money Amount	Fraction or Mixed Number	Decimal
6 pennies		$\frac{6}{100}$	0.06
	$0.50		0.50
		$\frac{70}{100}$ or $\frac{7}{10}$	0.70
3 $1 bills 9 pennies			3.09

Relate Fractions, Decimals, and Money

COMMON CORE STANDARD—4.NF.C.6
Understand decimal notation for fractions, and compare decimal fractions.

Write the total money amount. Then write the amount as a fraction or a mixed number and as a decimal in terms of dollars.

1.

$0.18; $\frac{18}{100}$; 0.18
_____

2.

_____

Write as a money amount and as a decimal in terms of dollars.

3. $\frac{25}{100}$ **4.** $\frac{79}{100}$ **5.** $\frac{31}{100}$ **6.** $\frac{8}{100}$ **7.** $\frac{42}{100}$

_____   _____   _____   _____   _____

Write the money amount as a fraction in terms of dollars.

8. $0.87   **9.** $0.03   **10.** $0.66   **11.** $0.95   **12.** $1.00

_____   _____   _____   _____   _____

Write the total money amount. Then write the amount as a fraction and as a decimal in terms of dollars.

13. 2 quarters 2 dimes   **14.** 3 dimes 4 pennies   **15.** 8 nickels 12 pennies

_____   _____   _____

Problem Solving *Real World*

16. Kate has 1 dime, 4 nickels, and 8 pennies. Write Kate's total amount as a fraction in terms of a dollar.

_____

17. **WRITE** *Math* Jeffrey says he has 6.8 dollars. How do you write the decimal 6.8 when it refers to money? Explain.

_____

© Houghton Mifflin Harcourt Publishing Company

Lesson Check (4.NF.C.6)

1. Write the total amount of money shown as a fraction in terms of a dollar.

_____

_____

2. Crystal has $\frac{81}{100}$ of a dollar. What could be the coins Crystal has?

_____

_____

Spiral Review (4.NF.A.1, 4.NF.C.6)

3. Joel gives $\frac{1}{3}$ of his baseball cards to his sister. Write a fraction that is equivalent to $\frac{1}{3}$.

_____

4. Penelope bakes pretzels. She salts $\frac{3}{8}$ of the pretzels. Write a fraction that is equivalent to $\frac{3}{8}$.

_____

5. What decimal is shown by the shaded area in the model?

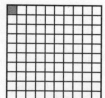

_____

6. Mr. Guzman has 100 cows on his dairy farm. Of the cows, 57 are Holstein. What decimal represents the portion of cows that are Holstein?

_____

FOR MORE PRACTICE
GO TO THE
Personal Math Trainer

Name _____

Problem Solving • Money

Essential Question How can you use the strategy *act it out* to solve problems that use money?

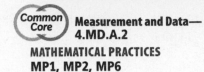

Common Core Measurement and Data—
4.MD.A.2
MATHEMATICAL PRACTICES
MP1, MP2, MP6

 Unlock the Problem Real World

Together, Marnie and Serena have $1.20. They want to share the money equally. How much money will each girl get?

Use the graphic organizer to solve the problem.

Read the Problem	Solve the Problem
What do I need to find?	You can make $1.20 with 4 quarters
I need to find the _____	and 2 _____.
_____	Circle the coins to show two sets with equal value.
What information do I need to use?	
I need to use the total amount, _____, and divide	
the amount into _____ equal parts.	
How will I use the information?	
I will use coins to model the _____ and	
act out the problem.	

So, each girl gets _____ quarters and

_____ dime. Each girl gets $_____.

• Describe another way you could act out the problem with coins.

_____

_____

🔑 Try Another Problem

Josh, Tom, and Chuck each have $0.40. How much money do they have together?

Read the Problem	Solve the Problem
What do I need to find?	
What information do I need to use?	
How will I use the information?	

- How can you solve the problem using dimes and nickels?

_____

_____

Math Talk

MATHEMATICAL PRACTICES ❶

Describe What other strategy might you use to solve the problem? Explain.

Name _____

Share and Show **MATH BOARD**

1. Juan has $3.43. He is buying a paint brush that costs $1.21 to paint a model race car. How much will Juan have after he pays for the paint brush?

First, use bills and coins to model $3.43.

Next, you need to subtract. Remove bills and coins that have a value of $1.21. Mark Xs to show what you remove.

Last, count the value of the bills and coins that are left. How much will Juan have left?

_____

2. What if Juan has $3.43, and he wants to buy a paint brush that costs $2.28? How much money will Juan have left then? Explain.

_____

_____

_____

_____

3. Sophia has $2.25. She wants to give an equal amount to each of her 3 young cousins. How much will each cousin receive?

_____

WRITE *Math*
Show Your Work

© Houghton Mifflin Harcourt Publishing Company

On Your Own

4. Marcus saves $13 each week. In how many weeks will he have saved at least $100?

_____

5. **MATHEMATICAL PRACTICE 1** **Analyze Relationships** Hoshi has $50. Emily has $23 more than Hoshi. Karl has $16 less than Emily. How much money do they have all together?

_____

6. **THINK SMARTER** Four girls have $5.00 to share equally. How much money will each girl get? Explain.

_____

_____

_____

WRITE *Math*
Show Your Work

7. **GO DEEPER** What if four girls want to share $5.52 equally? How much money will each girl get? Explain.

_____

_____

_____

Personal Math Trainer

8. **THINK SMARTER +** Aimee and three of her friends have three quarters and one nickel. If Aimee and her friends share the money equally, how much will each person get? Explain how you found your answer.

Problem Solving • Money

COMMON CORE STANDARD—4.MD.A.2
Solve problems involving measurement and conversion of measurements from a larger unit to a smaller unit.

Use the *act it out* strategy to solve.

1. Carl wants to buy a bicycle bell that costs $4.50.
   Carl has saved $2.75 so far. How much more
   money does he need to buy the bell?

   Use 4 $1 bills and 2 quarters to model $4.50.
   Remove bills and coins that have a value of $2.75.
   First, remove 2 $1 bills and 2 quarters.

   Next, exchange one $1 bill for 4 quarters and
   remove 1 quarter.

   Count the amount that is left.
   So, Carl needs to save $1.75 more.

   _____ $1.75

2. Together, Xavier, Yolanda, and Zachary have
   $4.44. If each person has the same amount, how
   much money does each person have?

   _____

3. Marcus, Nan, and Olive each have $1.65 in their
   pockets. They decide to combine the money. How
   much money do they have altogether?

   _____

4. Jessie saves $6 each week. In how many weeks
   will she have saved at least $50?

   _____

5. **WRITE** ▸*Math* Write a money problem you can solve
   using sharing, joining, or separating.

   _____

   _____

Lesson Check (4.MD.A.2)

1. Four friends earned $5.20 for washing a car. They shared the money equally. How much did each friend get?

2. Write a decimal that represents the value of one $1 bill and 5 quarters.

Spiral Review (4.OA.B.4, 4.NF.A.1, 4.NF.A.2, 4.NF.C.6)

3. Bethany has 9 pennies. What fraction of a dollar is this?

4. Michael made $\frac{9}{12}$ of his free throws at practice. What is $\frac{9}{12}$ in simplest form?

5. I am a prime number between 30 and 40. What number could I be?

6. Fill in the blank with a symbol that makes this statement true:

$$\frac{2}{5} \bigcirc \frac{1}{2}$$

FOR MORE PRACTICE
GO TO THE
Personal Math Trainer

Name _____

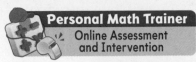

Personal Math Trainer
Online Assessment
and Intervention

Vocabulary

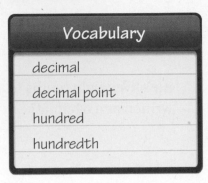

Vocabulary
decimal
decimal point
hundred
hundredth

Choose the best term from the box to complete the sentence.

1. A symbol used to separate the ones and the tenths place is

   called a _____ . (p. 495)

2. The number 0.4 is written as a _____ . (p. 495)

3. A _____ is one of one hundred equal parts of a

   whole. (p. 501)

Concepts and Skills

Write the fraction or mixed number and the decimal shown by the model. (4.NF.C.6)

4.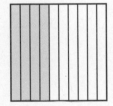

5.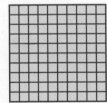

_____

_____   _____

Write the number as hundredths in fraction form and decimal form. (4.NF.C.5)

6. $\frac{8}{10}$

7. 0.5

8. $\frac{6}{10}$

_____   _____   _____

Write the fraction or mixed number as a money amount, and as a decimal in terms of dollars. (4.NF.C.6)

9. $\frac{65}{100}$

10. $1\frac{48}{100}$

11. $\frac{4}{100}$

_____   _____   _____

12. Ken's turtle competed in a 0.50-meter race. His turtle had traveled $\frac{49}{100}$ meter when the winning turtle crossed the finish line. What is $\frac{49}{100}$ written as a decimal? (4.NF.C.6)

_____

13. Alex lives eight tenths of a mile from Sarah. What is eight tenths written as a decimal? (4.NF.C.6)

_____

14. **GO DEEPER** What fraction and decimal, in hundredths, is equivalent to $\frac{7}{10}$? (4.NF.C.5)

_____

15. Elaine found the following in her pocket. How much money was in her pocket? (4.NF.C.6)

_____

16. Three girls share $0.60. Each girl gets the same amount. How much money does each girl get? (4.MD.A.2)

_____

17. The deli scale weighs meat and cheese in hundredths of a pound. Sam put $\frac{5}{10}$ pound of pepperoni on the deli scale. What weight does the deli scale show? (4.NF.C.5)

_____

Name _____

Add Fractional Parts of 10 and 100

Essential Question How can you add fractions when the denominators are 10 or 100?

Common Core Number and Operations—Fractions—4.NF.C.5 *Also 4.MD.A.2*
MATHEMATICAL PRACTICES
MP2, MP7, MP8

🔑 Unlock the Problem 🌎

The fourth grade classes are painting designs on tile squares to make a mural. Mrs. Kirk's class painted $\frac{3}{10}$ of the mural. Mr. Becker's class painted $\frac{21}{100}$ of the mural. What part of the mural is painted?

You know how to add fractions with parts that are the same size. You can use equivalent fractions to add fractions with parts that are not the same size.

🔒 Example 1 Find $\frac{3}{10} + \frac{21}{100}$.

STEP 1 Write $\frac{3}{10}$ and $\frac{21}{100}$ as a pair of fractions with a common denominator.

Think: 100 is a multiple of 10. Use 100 as the common denominator.

$$\frac{3}{10} = \frac{3 \times \boxed{}}{10 \times \boxed{}} = \frac{\boxed{}}{100}$$

Think: $\frac{21}{100}$ already has 100 in the denominator.

So, $\frac{\boxed{}}{100}$ of the mural is painted.

STEP 2 Add.

Think: Write $\frac{3}{10} + \frac{21}{100}$ using fractions with a common denominator.

$$\frac{30}{100} + \frac{21}{100} = \frac{\boxed{}}{100}$$

Math Talk

MATHEMATICAL PRACTICES ⑧

Draw Conclusions When adding tenths and hundredths, can you always use 100 as a common denominator? Explain.

Try This! Find $\frac{4}{100} + \frac{1}{10}$.

Ⓐ Write $\frac{1}{10}$ as $\frac{10}{100}$.

$$\frac{1}{10} = \frac{1 \times \boxed{}}{10 \times \boxed{}} = \frac{\boxed{}}{100}$$

Ⓑ Add.

$$\frac{\boxed{}}{100} + \frac{10}{100} = \frac{\boxed{}}{100}$$

So, $\frac{4}{100} + \frac{10}{100} = \frac{14}{100}$.

🔒 Example 2 Add decimals.

Sean lives 0.5 mile from the store. The store is 0.25 mile from his grandmother's house. Sean is going to walk to the store and then to his grandmother's house. How far will he walk?

Find 0.5 + 0.25.

STEP 1 Write 0.5 + 0.25 as a sum of fractions.

Think: 0.5 is 5 tenths.   Think: 0.25 is 25 hundredths.

$$0.5 = \frac{}{} \qquad\qquad 0.25 = \frac{}{}$$

Write 0.5 + 0.25 as $\dfrac{}{} + \dfrac{}{}$

STEP 2 Write $\frac{5}{10} + \frac{25}{100}$ as a sum of fractions with a common denominator.

Think: Use 100 as a common denominator. Rename $\frac{5}{10}$.

$$\frac{5}{10} = \frac{5 \times }{10 \times } = \frac{}{100}$$

Write $\frac{5}{10} + \frac{25}{100}$ as $\dfrac{}{} + \dfrac{}{}$.

STEP 3 Add.

$$\frac{50}{100} + \frac{25}{100} = \frac{}{}$$

STEP 4 Write the sum as a decimal.

So, Sean will walk _____ mile.

$$\frac{75}{100} = \underline{}$$

Math Talk

MATHEMATICAL PRACTICES ❼

Identify Relationships
Explain why you can think of $0.25 as either $\frac{1}{4}$ dollar or $\frac{25}{100}$ dollar.

Try This! Find $0.25 + $0.40.

$0.25 + $0.40 = _____

Remember
A money amount less than a dollar can be written as a fraction of a dollar.

Name _____

1. Find $\frac{7}{10} + \frac{5}{100}$.

   Think: Write the addends as fractions with a common denominator.

   $$\frac{}{100} + \frac{}{100} = \frac{}{}$$

Find the sum.

2. $\frac{1}{10} + \frac{11}{100} =$ _____

3. $\frac{36}{100} + \frac{5}{10} =$ _____

4. $0.16 + $0.45 = $ _____

5. $0.08 + $0.88 = $ _____

On Your Own

6. $\frac{6}{10} + \frac{25}{100} =$ _____

7. $\frac{7}{10} + \frac{7}{100} =$ _____

8. $0.55 + $0.23 = $ _____

9. $0.19 + $0.13 = $ _____

MATHEMATICAL PRACTICE ② **Reason Quantitatively** **Algebra** Write the number that makes the equation true.

10. $\frac{20}{100} + \frac{}{10} = \frac{60}{100}$

11. $\frac{2}{10} + \frac{}{100} = \frac{90}{100}$

12. GO DEEPER Jerry had 1 gallon of ice cream. He used $\frac{3}{10}$ gallon to make chocolate milkshakes and 0.40 gallon to make vanilla milkshakes. How much ice cream does Jerry have left after making the milkshakes?

_____

Problem Solving • Applications

Use the table for 13–16.

13. **THINK SMARTER** Dean selects Teakwood stones and Buckskin stones to pave a path in front of his house. How many meters long will each set of one Teakwood stone and one Buckskin stone be?

_____

Paving Stone Center	
Style	**Length (in meters)**
Rustic	$\frac{15}{100}$
Teakwood	$\frac{3}{10}$
Buckskin	$\frac{41}{100}$
Rainbow	$\frac{6}{10}$
Rose	$\frac{8}{100}$

14. The backyard patio at Nona's house is made from a repeating pattern of one Rose stone and one Rainbow stone. How many meters long is each pair of stones?

_____

15. **GO DEEPER** For a stone path, Emily likes the look of a Rustic stone, then a Rainbow stone, and then another Rustic stone. How long will the three stones in a row be? Explain.

_____

_____

16. **WRITE** ›Math Which two stones can you place end-to-end to get a length of 0.38 meter? Explain how you found your answer.

_____

_____

17. **THINK SMARTER** Christelle is making a dollhouse. The dollhouse is $\frac{6}{10}$ meter tall without the roof. The roof is $\frac{15}{100}$ meter high. What is the height of the dollhouse with the roof? Choose a number from each column to complete an equation to solve.

$$\frac{6}{10} + \frac{15}{100} = \boxed{\begin{array}{c}\frac{6}{100}\\[4pt]\frac{60}{100}\\[4pt]\frac{61}{100}\end{array}} + \boxed{\begin{array}{c}\frac{15}{10}\\[4pt]\frac{5}{100}\\[4pt]\frac{15}{100}\end{array}} = \boxed{\begin{array}{c}\frac{65}{100}\\[4pt]\frac{7}{10}\\[4pt]\frac{75}{100}\end{array}} \text{ meter.}$$

Add Fractional Parts of 10 and 100

COMMON CORE STANDARD—4.NF.C.5
*Understand decimal notation for fractions,
and compare decimal fractions.*

Find the sum.

1. $\dfrac{2}{10} + \dfrac{43}{100}$

    $\dfrac{20}{100} + \dfrac{43}{100} = \dfrac{63}{100}$

    $\dfrac{63}{100}$

Think: Write $\dfrac{2}{10}$ as a fraction with a denominator of 100:

$\dfrac{2 \times 10}{10 \times 10} = \dfrac{20}{100}$

2. $\dfrac{17}{100} + \dfrac{6}{10}$

3. $\dfrac{9}{100} + \dfrac{9}{10}$

4. $\$0.25 + \$0.34$

Problem Solving · Real World

5. Ned's frog jumped $\dfrac{38}{100}$ meter. Then his frog jumped $\dfrac{4}{10}$ meter. How far did Ned's frog jump?

6. Keiko walks $\dfrac{5}{10}$ kilometer from school to the park. Then she walks $\dfrac{19}{100}$ kilometer from the park to her home. How far does Keiko walk?

7. **WRITE** ▸*Math* Explain how you would use equivalent fractions to solve $0.5 + 0.10$.

_____

_____

_____

_____

Lesson Check (4.NF.C.5)

1. In a fish tank, $\frac{2}{10}$ of the fish were orange and $\frac{5}{100}$ of the fish were striped. What fraction of the fish were orange or striped?

2. Greg spends $0.45 on an eraser and $0.30 on a pen. How much money does Greg spend?

Spiral Review (4.NF.A.1, 4.NF.B.3d, 4.MD.A.2)

3. Phillip saves $8 each month. How many months will it take him to save at least $60?

4. Ursula and Yi share a submarine sandwich. Ursula eats $\frac{2}{8}$ of the sandwich. Yi eats $\frac{3}{8}$ of the sandwich. How much of the sandwich do the two friends eat?

5. A carpenter has a board that is 8 feet long. He cuts off two pieces. One piece is $3\frac{1}{2}$ feet long and the other is $2\frac{1}{3}$ feet long. How much of the board is left?

6. Jeff drinks $\frac{2}{3}$ of a glass of juice. Write a fraction that is equivalent to $\frac{2}{3}$.

© Houghton Mifflin Harcourt Publishing Company

FOR MORE PRACTICE
GO TO THE
Personal Math Trainer

Name _____

Compare Decimals

Essential Question How can you compare decimals?

 Common Core Number and Operations—Fractions—4.NF.C.7
MATHEMATICAL PRACTICES
MP2, MP6, MP7

Unlock the Problem

The city park covers 0.64 square mile. About 0.18 of the park is covered by water, and about 0.2 of the park is covered by paved walkways. Is more of the park covered by water or paved walkways?

- Cross out unnecessary information.
- Circle numbers you will use.
- What do you need to find?

_____

_____

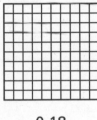 One Way Use a model.

Shade 0.18.   Shade 0.2.

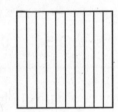

0.18   ◯   0.2

Other Ways

A Use a number line.

Locate 0.18 and 0.2 on a number line.

Think: 2 tenths is equivalent to 20 hundredths.

0.0   0.10   0.20   0.30   0.40   0.50

_____ is closer to 0, so 0.18 ◯ 0.2.

Math Talk

MATHEMATICAL PRACTICES ➏

Compare How does the number of tenths in 0.18 compare to the number of tenths in 0.2? Explain.

B Compare equal-size parts.

- 0.18 is _____ hundredths.

- 0.2 is 2 tenths, which is equivalent to _____ hundredths.

18 hundredths ◯ 20 hundredths, so 0.18 ◯ 0.2.

So, more of the park is covered by _____.

Place Value You can compare numbers written as decimals by using place value. Comparing decimals is like comparing whole numbers. Always compare the digits in the greatest place-value position first.

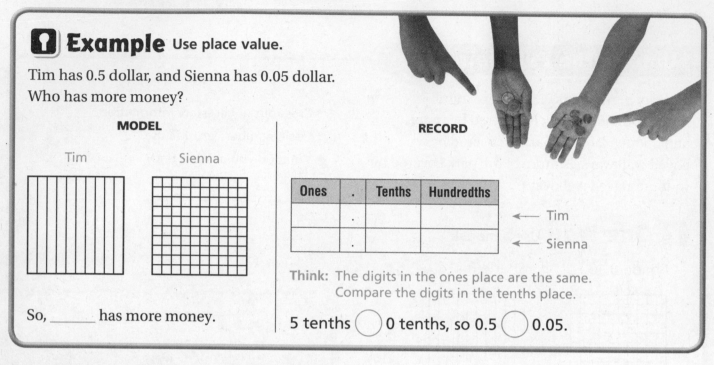

🔓 **Example** Use place value.

Tim has 0.5 dollar, and Sienna has 0.05 dollar. Who has more money?

MODEL

Tim   Sienna

So, _____ has more money.

RECORD

Ones	.	Tenths	Hundredths
	.			← Tim
	.			← Sienna

Think: The digits in the ones place are the same. Compare the digits in the tenths place.

5 tenths ◯ 0 tenths, so 0.5 ◯ 0.05.

- Compare the size of 1 tenth to the size of 1 hundredth. How could this help you compare 0.5 and 0.05? Explain.

_____

_____

_____

Try This! Compare 1.3 and 0.6. Write <, >, or =.

1.3 ◯ 0.6

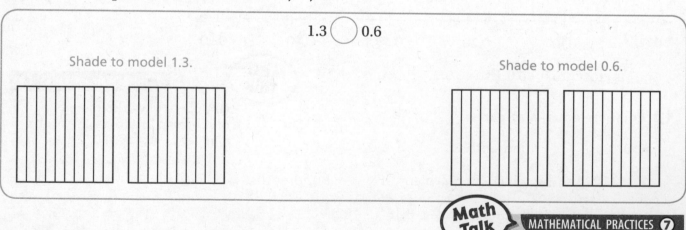

Shade to model 1.3.

Shade to model 0.6.

Math Talk

© Houghton Mifflin Harcourt Publishing Company

MATHEMATICAL PRACTICES ❼

Look for Structure How could you use place value to compare 1.3 and 0.6?

Name _____

1. Compare 0.39 and 0.42. Write <, >, or =.
   Shade the model to help.

   0.39 ◯ 0.42

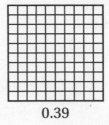

0.39

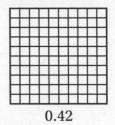

0.42

Compare. Write <, >, or =.

2. 0.26 ◯ 0.23

Ones	.	Tenths	Hundredths
	.		
	.		

✓ 3. 0.7 ◯ 0.54

Ones	.	Tenths	Hundredths
	.		
	.		

4. 1.15 ◯ 1.3

Ones	.	Tenths	Hundredths
	.		
	.		

✓ 5. 4.5 ◯ 2.89

Ones	.	Tenths	Hundredths
	.		
	.		

Math Talk

MATHEMATICAL PRACTICES ②

Reason Abstractly Can you compare 0.39 and 0.42 by comparing only the tenths? Explain.

On Your Own

Compare. Write <, >, or =.

6. 0.9 ◯ 0.81

7. 1.06 ◯ 0.6

8. 0.25 ◯ 0.3

9. 2.61 ◯ 3.29

MATHEMATICAL PRACTICE ② **Reason Quantitatively** Compare. Write <, >, or =.

10. 0.30 ◯ $\frac{3}{10}$

11. $\frac{4}{100}$ ◯ 0.2

12. 0.15 ◯ $\frac{1}{10}$

13. $\frac{1}{8}$ ◯ 0.8

14. **GO DEEPER** Robert had $14.53 in his pocket. Ivan had $14.25 in his pocket. Matt had $14.40 in his pocket. Who had more money, Robert or Matt? Did Ivan have more money than either Robert or Matt?

Unlock the Problem Real World

15. **THINK SMARTER** Ricardo and Brandon ran a 1500-meter race. Ricardo finished in 4.89 minutes. Brandon finished in 4.83 minutes. What was the time of the runner who finished first?

a. What are you asked to find? _____

b. What do you need to do to find the answer? _____

_____

c. Solve the problem.

d. What was the time of the runner who finished first?

_____

e. Look back. Does your answer make sense? Explain.

_____

_____

_____

_____

16. **GO DEEPER** The Venus flytrap closes in 0.3 second and the waterwheel plant closes in 0.2 second. What decimal is halfway between 0.2 and 0.3? Explain.

_____

_____

_____

_____

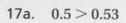
Personal Math Trainer

17. **THINK SMARTER +** For numbers 17a–17c, compare then select True or False.

17a. $0.5 > 0.53$ ○ True   ○ False

17b. $0.35 < 0.37$ ○ True   ○ False

17c. $\$1.35 > \$0.35$ ○ True   ○ False

Name _____

Compare Decimals

COMMON CORE STANDARDS—4.NF.C.7
Understand decimal notation for fractions, and compare decimal fractions.

Compare. Write <, >, or =.

1. 0.35 ⟨ < ⟩ 0.53

Think: 3 tenths is less
than 5 tenths.
So, 0.35 < 0.53

2. 0.6 ◯ 0.60

3. 0.24 ◯ 0.31

4. 0.94 ◯ 0.9

5. 0.3 ◯ 0.32

6. 0.45 ◯ 0.28

7. 0.39 ◯ 0.93

Use the number line to compare. Write *true* or *false*.

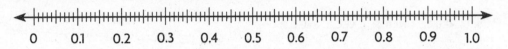

8. 0.8 > 0.78

_____

9. 0.4 > 0.84

_____

10. 0.7 < 0.70

_____

11. 0.4 > 0.04

_____

Compare. Write *true* or *false*.

12. 0.09 > 0.1

_____

13. 0.24 = 0.42

_____

14. 0.17 < 0.32

_____

15. 0.85 > 0.82

_____

Problem Solving *Real World*

16. Kelly walks 0.7 mile to school. Mary walks 0.49 mile to school. Write an inequality using <, >, or = to compare the distances they walk to school.

_____

_____

17. **WRITE** *Math* Show or describe two different ways to complete the comparison using <, >, or =: 0.26 ◯ 0.4.

_____

_____

_____

Lesson Check (4.NF.C.7)

1. Bob, Cal, and Pete each made a stack of baseball cards. Bob's stack was 0.2 meter high. Cal's stack was 0.24 meter high. Pete's stack was 0.18 meter high. Write a number sentence that compares Cal's stack of cards to Pete's stack of cards.

2. Three classmates spent money at the school supplies store. Mark spent 0.5 dollar, Andre spent 0.45 dollar, and Raquel spent 0.52 dollar. Write a number sentence that compares the money Andre spent to the money that Mark spent.

Spiral Review (4.NF.B.3c, 4.NF.B.4c, 4.NF.C.5, 4.NF.C.6)

3. Pedro has $0.35 in his pocket. Alice has $0.40 in her pocket. How much money do Pedro and Alice have altogether?

4. The measure 62 centimeters is equivalent to $\frac{62}{100}$ meter. What is this measure written as a decimal?

5. Joel has 24 sports trophies. Of the trophies, $\frac{1}{8}$ are soccer trophies. How many soccer trophies does Joel have?

6. Molly's jump rope is $6\frac{1}{3}$ feet long. Gail's jump rope is $4\frac{2}{3}$ feet long. How much longer is Molly's jump rope?

FOR MORE PRACTICE
GO TO THE
Personal Math Trainer

✓ Chapter 9 Review/Test

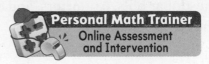

Personal Math Trainer
Online Assessment and Intervention

1. Select a number shown by the model. Mark all that apply.

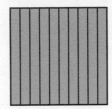 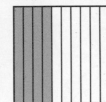

$\frac{14}{10}$	$\frac{40}{10}$	1.4
$1\frac{4}{10}$	14	4.1

2. Rick has one dollar and twenty-seven cents to buy a notebook. Which names this money amount in terms of dollars? Mark all that apply.

   (A) 12.7   (D) 1.27

   (B) 1.027   (E) $1\frac{27}{100}$

   (C) $1.27   (F) $\frac{127}{10}$

3. For numbers 3a–3e, select True or False for the statement.

   3a. 0.9 is equivalent to 0.90.   ○ True   ○ False

   3b. 0.20 is equivalent to $\frac{2}{100}$.   ○ True   ○ False

   3c. $\frac{80}{100}$ is equivalent to $\frac{8}{10}$.   ○ True   ○ False

   3d. $\frac{6}{10}$ is equivalent to 0.60.   ○ True   ○ False

   3e. 0.3 is equivalent to $\frac{3}{100}$.   ○ True   ○ False

4. After selling some old books and toys, Gwen and her brother Max had 5 one-dollar bills, 6 quarters, and 8 dimes. They agreed to divide the money equally.

Part A

What is the total amount of money that Gwen and Max earned? Explain.

```
┌─────────────────────────────────────────────────┐
│                                                   │
│                                                   │
│                                                   │
│                                                   │
│                                                   │
└─────────────────────────────────────────────────┘
```

Part B

Max said that he and Gwen cannot get equal amounts of money because 5 one-dollar bills cannot be divided evenly. Do you agree with Max? Explain.

```
┌─────────────────────────────────────────────────┐
│                                                   │
│                                                   │
│                                                   │
│                                                   │
│                                                   │
│                                                   │
└─────────────────────────────────────────────────┘
```

5. Harrison rode his bike $\frac{6}{10}$ of a mile to the park. Shade the model. Then write the decimal to show how far Harrison rode his bike.

```
┌──┬──┬──┬──┬──┬──┬──┬──┬──┬──┐
│  │  │  │  │  │  │  │  │  │  │
└──┴──┴──┴──┴──┴──┴──┴──┴──┴──┘
```

Harrison rode his bike _____ mile to the park.

6. Amaldo spent $\frac{88}{100}$ of a dollar on a souvenir pencil from Zion National Park in Utah. What is $\frac{88}{100}$ written as a decimal in terms of dollars?

```
┌──────┐
│      │
└──────┘
```

7. Tran has $5.82. He is saving for a video game that costs $8.95.

Tran needs _____ more to have enough money for the game.

540

8. Cheyenne lives $\frac{7}{10}$ mile from school. A fraction in hundredths equal to $\frac{7}{10}$ is _____.

9. Write a decimal in tenths that is **less** than 2.42 but **greater** than 2.0.

_____

10. GO DEEPER Kylee and two of her friends are at a museum. They find two quarters and one dime on the ground.

Part A

If Kylee and her friends share the money equally, how much will each person get? Explain how you found your answer.

Part B

Kylee says that each person will receive $\frac{2}{10}$ of the money that was found. Do you agree? Explain.

11. Shade the model to show $1\frac{52}{100}$. Then write the mixed number in decimal form.

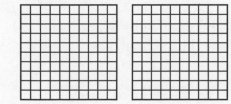

12. Henry is making a recipe for biscuits. A recipe calls for $\frac{5}{10}$ kilogram flour and $\frac{9}{100}$ kilogram sugar.

Part A

If Henry measures correctly and combines the two amounts, how much flour and sugar will he have? Show your work.

Part B

How can you write your answer as a decimal?

13. An orchestra has 100 musicians. $\frac{40}{100}$ of them play string instruments—violin, viola, cello, double bass, guitar, lute, and harp. What decimal is equivalent to $\frac{40}{100}$?

14. Complete the table.

$ Bills and Coins	Money Amount	Fraction or Mixed Number	Decimal
8 pennies		$\frac{8}{100}$	0.08
	$0.50		0.50
		$\frac{90}{100}$ or $\frac{9}{10}$	0.90
4 $1 bills 5 pennies			4.05

15. The point on the number line shows the number of seconds it took an athlete to run the forty-yard dash. Write the decimal that correctly names the point.

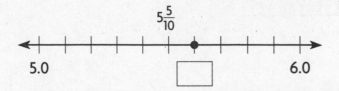

542

Name _____

16. Ingrid is making a toy car. The toy car is $\frac{5}{10}$ meter high without the roof. The roof is $\frac{18}{100}$ meter high. What is the height of the toy car with the roof? Choose a number from each column to complete an equation to solve.

$$\frac{5}{10} + \frac{18}{100} = \boxed{\begin{array}{c} \frac{5}{100} \\ \frac{15}{100} \\ \frac{50}{100} \end{array}} + \boxed{\begin{array}{c} \frac{18}{100} \\ \frac{81}{100} \\ \frac{18}{10} \end{array}} = \boxed{\begin{array}{c} \frac{68}{10} \\ \frac{32}{100} \\ \frac{68}{100} \end{array}} \text{ meter high.}$$

17. Callie shaded the model to represent the questions she answered correctly on a test. What decimal represents the part of the model that is shaded?

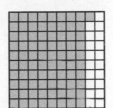

represents [   ]

18. THINK SMARTER + For numbers 18a–18f, compare then select True or False.

18a.   0.21 < 0.27   ○ True   ○ False

18b.   0.4 > 0.45   ○ True   ○ False

18c. $3.21 > $0.2   ○ True   ○ False

18d.   1.9 < 1.90   ○ True   ○ False

18e.   0.41 = 0.14   ○ True   ○ False

18f.   6.2 > 6.02   ○ True   ○ False

19. Fill in the numbers to find the sum.

$$\frac{4}{10} + \frac{\boxed{}}{100} = \frac{8}{\boxed{}}$$

20. Steve is measuring the growth of a tree. He drew this model to show the tree's growth in meters. Which fraction, mixed number, or decimal does the model show? Mark all that apply.

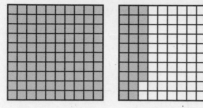

(A) 1.28

(B) 12.8

(C) 0.28

(D) $2\frac{8}{100}$

(E) $1\frac{28}{100}$

(F) $1\frac{28}{10}$

21. Luke lives 0.4 kilometer from a skating rink. Mark lives 0.25 kilometer from the skating rink.

Part A

Who lives closer to the skating rink? Explain.

Part B

How can you write each distance as a fraction? Explain.

Part C

Luke is walking to the skating rink to pick up a practice schedule. Then he is walking to Mark's house. Will he walk more than a kilometer or less than a kilometer? Explain.